Bernardo Carvalhaes

Eco-efficiency: Analysing Atmospheric Emissions from a Railway

Bernardo Carvalhaes

Eco-efficiency: Analysing Atmospheric Emissions from a Railway

Analysing Atmospheric Emissions from a Railway

ScienciaScripts

Cover image: www.ingimage.com

This book is a translation from the original published under ISBN 978-613-9-69669-7.

Publisher:
Sciencia Scripts
is a trademark of
Dodo Books Indian Ocean Ltd. and OmniScriptum S.R.L publishing group

120 High Road, East Finchley, London, N2 9ED, United Kingdom
Str. Armeneasca 28/1, office 1, Chisinau MD-2012, Republic of Moldova, Europe
Printed at: see last page
ISBN: 978-620-8-21075-5

SUMMARY

CHAPTER 1

INTRODUCTION

The good performance of the railway sector in recent years, coupled with the expectation that the volumes transported will continue to increase each year, means that more and more attention is being paid to this important mode of transport. According to the National Association of Rail Transporters (ANTF), the volume of goods transported by train jumped from 308 million tonnes in 2014 to 332 million tonnes in 2015, with a similar increase expected for 2016 (ANTF, 2014). Most of the cargo transported by rail in the country consists of iron ore and coal, accounting for approximately 75 per cent of the total. Even though it is a mode of transport with low atmospheric emissions, much lower than road transport, for example, this huge volume of goods transported means that thousands of tonnes of pollutants are emitted into the atmosphere.

More than a fifth of this volume is transported by the Vitória-Minas Railway (EFVM), located in the south-east of the country. With a length of 905 kilometres, the railway transports approximately 119 million tonnes of iron ore every year, as well as more than 20 million tonnes of general cargo (soya, corn, fertilisers, coal, steel, etc.). Its fleet consists of more than 300 locomotives and 14,000 wagons. It is no coincidence that this railway was selected to apply the method proposed in the book, and the results will be shown throughout the text.

The aim of this book was to analyse the atmospheric emissions caused by the operation of locomotives on a railway, obtain their eco-efficiency and offer suggestions for alternative fuels and technologies to reduce pollution. This was made possible using a method designed to assess atmospheric emissions, energy efficiency and costs arising from the use of different fuels and to compare their performance with the standard diesel oil used by locomotives, which is made up of 5% biodiesel and 95% diesel oil, also known as B5. The method used is an evolution from traditional energy efficiency programmes - which focus on reducing consumption - to a new approach that evaluates the cost/benefit of improving the economic and environmental aspects of the railway business, and can be applied to railways all over the world and of all types - freight or passenger.

CHAPTER 2

ECO-EFFICIENCY

The concept of eco-efficiency arose to show that it is possible to unite the - apparently antagonistic - ideas of financial performance and environmental preservation. Talks on this subject began when several companies suggested working together to focus on the environment during the United Nations Conference on Environment and Development (ECO-92), which took place in Rio de Janeiro in 1992 (LEAL JUNIOR, 2010).

At the time, a competition was held to find ideas for implementing sustainable development in companies, materialising the concept into something practical and immediate, and the winning idea was eco-efficiency, proposed by the World Business Council for Sustainable Development (WBCSD) (WBCSD, 2000). The need to produce more with fewer resources, less waste and less pollution already existed, but the methodology proposed by the WBCSD referred to ecology, economy and efficiency in an innovative way until then (HOLLIDAY et al. 2002).

The WBCSD is a group of 130 international companies that share the same commitment: sustainable development. The network of members that make up the coalition involves more than 30 countries and 20 key sectors of world industry, all pursuing the same goal: to act as generators of change, stimulating co-operation between public administration and the private sector. The main areas that receive attention from the methodology are (WBCSD, 2000):

- Communicating sustainability;
- Sustainability through the market;
- Innovation and development of new technologies;
- Climate and energy studies;

- Preservation of natural resources.

Eco-efficiency can be defined as "the delivery of goods and services at competitive prices, which at the same time satisfy human needs, result in a better quality of life and gradually reduce environmental influences as well as the intensity of resource use at all stages of the production cycle of a good or service" (WBCSD, 2000) and has three main objectives:

1) Reduction in the consumption of natural inputs, through less use of natural resources of a material and energy nature, concern for recycling and the durability of products.
2) Reduction of factors that negatively impact the environment, such as the emission of polluting gases, liquid discharges and the spread of possible toxic substances;
3) Increasing the value of the product or service, generating products with greater added value, thus benefiting end customers.

The company interested in applying the method must select indicators so that eco-efficiency can be measured, which must follow the principles below:

1) Being relevant to protecting the environment, human health and improving quality of life: the indicators should ensure that companies and stakeholders focus their efforts on priority areas, both in terms of the environment and adding value to the product;
2) Providing information to decision-making bodies in order to improve the organisation's performance: the primary function of indicators is to help management make decisions regarding changes to production processes, reducing the use of resources and/or adding value to the product;
3) Recognise the diversity present in each business: the differences existing in each type of business must be taken into account in order to select indicators that are important and meaningful, so as to be able to portray reality;

4) Support benchmarking and monitor evolution: it is essential that management uses indicators whose evolution can be measured historically. In order to maximise monitoring power, indicators should be calculated to minimise the influence of external factors that affect both the environmental impact and the value of the product. In addition, the indicators must be based on solid, timeless and reproducible measurement systems;
5) Clearly defined, measurable, transparent and verifiable: in order to support decision-making, indicators must be clear and calculated accurately or by reliable estimates. Therefore, the entire calculation process and the considerations made must be documented and made available to employees;
6) Be understandable and meaningful to the various stakeholders: the indicators should be simple and clear, so as to make them easy to use for both the company and the other stakeholders. It is important to emphasise that any aggregation of data must be done clearly and carefully, making it clear what limitations have been found;
7) Based on a general assessment of the company's activity (products and/or services), above all concentrating on those areas directly controlled by management: the entire company should be examined in search of a comprehensive understanding, prioritising the creation of indicators for areas in which management has direct control and can influence more quickly and precisely;
8) Taking into account relevant and significant issues related to the company's upstream activities (suppliers) and downstream activities (customers): management should not overlook areas of the company over which it does not have full control (e.g. production of raw materials), but the treatment should be differentiated from those indicators fully controlled by management.

2.1 MEASURING ECO-EFFICIENCY

Measuring eco-efficiency allows companies to monitor their activities, as well as control and set short- and long-term targets. Control must be carried out by all employees and publicity must be broad, clear and simple, involving the entire business (PEREIRA, 2001).

Measuring eco-efficiency as proposed by the WBCSD is done using the formula:

$$\text{Eco-efficiency} = \frac{\text{Value of the product or service}}{\text{Environmental influence}} \qquad (1)$$

(unit: to be defined for each indicator)

There is no definition of what the numerators (value of the product or service) and denominators (environmental influence) of the above formula should be (PEREIRA, 2001), but the WBCSD suggests some indicators based on studies already carried out by companies in various parts of the world, allowing for a wide range of options according to the needs of the business. Even so, in rail transport no specific set of numerators and denominators has yet been proposed - this is done later in the book.

2.1.1 Type of indicators

There are two types of indicators: 1) generally applicable and 2) specific to the business. Generally applicable indicators (Table 1) are those that can be used in practically any type of company, industry and/or organisation in order to focus on a particular environmental concern or business value (WBCSD, 2000).

Table 1 - Examples of generally applicable indicators

Type	Indicator	Service
Product/service value	Quantity of product produced	Responsible for measuring the product manufactured within a time horizon
	Number of services provided	Responsible for measuring the amount of service provided within a time horizon
	Net sales	Total net sales (excluding costs)
Environmental influence	Energy consumption	Measuring the consumption of different energy sources
	Raw material consumption	Measuring the consumption of the various raw materials required
	Water consumption	Measuring the consumption of the various raw materials required
	Greenhouse gas emissions	Measurement of the gases produced in each product
	Emissions of substances that deplete the ozone layer (SDCO)	Measurement of the gases produced in each product

Source: WBCSD (2000).

Business-specific indicators, on the other hand, are those that are geared to the particular characteristics of each company, and each one must assess its business and identify which indicators should be studied.

2.1.2 Category, aspect and indicators

In order to obtain a methodology for creating and measuring universal indicators for all types of business, not excluding the particularities of each sector, the WBCSD (2000) determined a methodology consistent with those used in the ISO 14000 series and the Global Reporting Initiative (GRI), encompassing three organisational levels:

1. **Categories**: these are the areas in which there is an environmental influence (both on the creation and use of goods or services) and on the value of the business;
2. **Aspects**: each category has a certain number of aspects that are related to each other, with the aim of describing what will be measured;
3. **Indicators**: are responsible for measuring an individual aspect, which has several indicators.

To summarise, the main aspects identified within each **category** are (WBCSD, 2000):

1) Value of the product or service:
    - Volume/mass;
    - Monetary value;
    - Role.
2) Environmental influence on the creation of the product or service:
    - Energy consumption;
    - Consumption of materials;
    - Consumption of natural resources;
    - Unforeseen events.
3) Environmental influence on the use of the product or service:
    - Characteristics of the product or service;

- Packaging waste;
- Energy consumption;
- Emissions during use/emission.

Based on the formula for calculating eco-efficiency proposed by the WBCSD, Equation (1), a method for calculating eco-efficiency adapted for locomotives is proposed below.

CHAPTER 3

CALCULATING ECO-EFFICIENCY IN RAILWAYS

This chapter explains how to calculate eco-efficiency in railways using the WBCSD formula.

3.1 METHOD OF CALCULATING INDICATORS

To calculate the eco-efficiency indicators, it is necessary to define the service value (V) and the values of the eco-efficiency measures (ME). For the service value (V), an indicator used worldwide by railway operators called Tonnes per Gross Kilometre (TKB) was chosen. It represents the sum of the weight of the train (payload plus the weight of the wagons and locomotives) multiplied by the distance transported.

For TKB to be obtained, you need 1) the number of wagons (W) and 2) the number of journeys per year made by each wagon (T). In the following formula, T_{wt} is the gross tonnes (dead weight plus cargo weight) of each wagon $w \in W$ on each trip $t \in T$ (tonnes) and K_{wt} is the total number of kilometres covered by each wagon $w \in W$ during the trip $t \in T$ (km).

$$TKB = \sum_{w \in W} \sum_{t \in T} TB_{wt} K_{wt} \qquad (2)$$

(unit: ton x km)

Seven eco-efficiency (EM) measures were then defined and are presented below:

1) Total energy consumption;

2) Total consumption of renewable energy;
3) Carbon dioxide emissions;

4) Carbon monoxide emissions;

5) Nitrogen oxide emissions;

6) Particulate matter emissions;

7) Total cost (investment, maintenance and fuel).

3.1.1 Environmental Influence Indicators

To calculate these seven measures, the seven environmental influence indicators (EI) must first be calculated, for which "F" represents the combination of fuels that can be used by the locomotives and "L" the combination of locomotives.

In order to calculate total energy consumption (eiE), the following values need to be calculated: FEfl, which represents the consumption of fuel f∈ F by locomotive l G L (l/h); WH_{fl} , which are the hours worked by locomotive l∈ L using fuel f∈ F (h) and CCf, which is the calorific value of fuel f∈ F (MJ/l). Therefore, the total energy consumption can be obtained from the formula:

$$eiE = \sum_{f \in F} CC_f \left(\sum_{l \in L} FE_{fl} WH_{fl} \right) \qquad (3)$$

(unit: MJ)

Total Renewable Energy Consumption (eiRE) is determined using the formula below, but only takes into account locomotives using renewable fuels (F^r), such as biodiesel.

$$eiRE = \sum_{f \in F^r} CC_f \left(\sum_{l \in L^r} FE_{fl} WH_{fl} \right) \qquad (4)$$

(unit: MJ)

The total carbon dioxide emission (ei_{CO2}) represents the total carbon dioxide emissions in kilograms (kg) into the atmosphere, where EFf is the CO_2 emission factor of the fuel f∈ F ($kgCO_2/l$ or kgCO2/m).[3]

$$eiCO_2 = \sum_{f\in F} EF_f \left(\sum_{l\in L} FE_{fl} WH_{fl} \right) \quad (5)$$

(unit: $kgCO_2$)

The Carbon Monoxide (eiCO), Nitrogen Oxide ($eiNO_X$) and Particulate Matter (eiPM) emission indicators are measured in grams per litre or grams per cubic metre for LNG. The formulas below differ from each other by the emission factor for each pollutant, respectively EO_f, EN_f and EP_f.

$$eiCO = \sum_{f\in F} EO_f \left(\sum_{l\in L} FE_{fl} WH_{fl} \right) \quad (6)$$

(unit: gCO)

$$eiNO_x = \sum_{f\in F} EN_f \left(\sum_{l\in L} FE_{fl} WH_{fl} \right) \quad (7)$$

(unit: gNO_X)

$$eiPM = \sum_{f\in F} EP_f \left(\sum_{l\in L} FE_{fl} WH_{fl} \right) \quad (8)$$

(unit: kgPM)

The last indicator is Cost Efficiency (eiCE) and represents the total cost needed to operate with a given fuel, calculated as the sum of the cost of fuel, maintenance and investment. The fuel cost (FC) represents the total cost (US$) of purchasing the fuel consumed by the railway's locomotives. It is calculated using the formula below in which FU_f represents the cost of fuel $f \in F$ (US$ / lou US$ / m).[3]

$$FC = \sum_{f\in F} FU_f \left(\sum_{l\in L} FE_{fl} WH_{fl} \right) \quad (9)$$

(unit: US$)

When operating with an alternative fuel, additional maintenance is required to repair the dual-fuel kits installed in diesel engines in the case of LNG operation. In the case of biodiesel, additional maintenance is required for various components during operation. The cost of additional maintenance (MC) is determined using the formula below, where MUf is the additional cost of maintenance per hour worked (US$/h).

$$MC = \sum_{f \in F} MU_f \left(\sum_{l \in L} WH_{fl} \right) \quad (10)$$

(unit: US$)

Some investments are necessary to operate on natural gas. Therefore, the capital expenditure (CAPEX) for installing dual-fuel kits in locomotives needs to be taken into account. No CAPEX is required for operation with B25. The investment cost (IC) is calculated using the following formula, where IU_f represents the total investment when using fuel $f \in F$ (US$).

$$IC = \sum_{f \in F} IU_f \quad (11)$$

(unit: US$)

Once all the costs have been determined, the cost efficiency indicator (US$) eiCE can then be calculated from the fuel cost (FC), maintenance cost (MC) and investment cost (IC). This leads to the following formula, by which the cost efficiency eiCE (other environmental influence) in US$ can be determined.

$$eiCE = FC + MC + IC \quad (12)$$

(unit: US$)

3.1.2 Eco-efficiency measures

Once all the environmental influence and service value indicators have been obtained, it is possible to calculate the seven Eco-efficiency Measures listed in the formulae below: 1) total energy consumption (E), 2) total renewable energy consumption (RE), 3) carbon dioxide emissions (CO_2), 4) carbon monoxide emissions (CO), 5) nitrogen oxides emissions (NO_x), 6) particulate matter emissions (PM) and 7) cost efficiency (investment, maintenance and fuels) (CE).

E = TKB / eiE (unit:tonxkm/MJ) (13)
RE = TKB / eiRE (unit:tonxkm/MJ) (14)
CO_2 = TKB / eiCO_2 (unit:tonxkm/kgCO_2) (15)
CO = TKB / eiCO (unit:tonxkm/gCO) (16)

$NO_X = TKB / eiNO_X$ (unit:tonxkm/gNO_X)(17)
$PM = TKB / eiPM$ (unit:tonxkm/gPM) (18)
$CE = TKB / eiCE$ (unit:tonxkm/US$) (19)

These Eco-efficiency Measures are therefore the result of adapting Equation 1 for use on railways, with Eco-efficiency being the result of dividing the TKB (value of the service, "V") by the quantity of each pollutant emitted, NO_X, CO, CO_2 and PM. In the case of the Energy Consumption (E) and Renewable Energy Consumption (RE) equations, the denominators of Equation 1 are measures of energy consumption, in Megajoules, with the TKB also being the numerator. Finally, there is the cost as another environmental influence, considered in US$, which is the denominator of Equation 1 together with TKB as the numerator. Equations 13 to 19 can therefore be interpreted as a re-reading of Equation 1, proposed by the WBCSD, as shown in the following example for the case of Energy Consumption (E):

$$\textit{Eco-efficiency (E)} = \frac{\textit{Value of the product or service (TKB)}}{\textit{Environmental influence (MJ)}} \qquad (20)$$

(unit: ton x km/MJ)

In the case of pollutant emissions, we would have the following reinterpretation, exemplified by the pollutant CO_2 and which follows the same logic for the other pollutants, but measured in grams instead of kilograms:

$$\textit{Eco-efficiency } (C0_2) = \frac{\textit{Value of the product or service (TKB)}}{\textit{Environmental influence } (kgC0\,)_2} \qquad (21)$$

(unit: ton x km/kgCO_2)

Finally, to evaluate the cost, we would have Equation 1 represented as follows:

$$\textit{Eco-efficiency (EC)} = \frac{\textit{Value of the product or service (TKB)}}{\textit{Environmental influence (US\$)}} \qquad (22)$$

(unit: ton x km/US$)

For all indicators, higher values indicate better results, since for the same value of service (TKB) the railway uses the energy source more efficiently, emits fewer pollutants or spends fewer financial resources to switch to a cleaner fuel.

Table 2 therefore shows the eco-efficiency measures and their objectives, as well as listing the equations for determining each one.

Table 2 - Eco-efficiency measures, objectives and proposed equations

Eco-efficiency Measure (EM)	Objective	Equation
Energy efficiency	Evaluate fuel consumption	(13)
Renewable energy efficiency	Evaluate the use of renewable energy	(14)
Emission efficiency CO2	Evaluating CO2 emissions	(15)
Emission efficiency CO	Evaluating CO	(16)
Emission efficiency MP	Evaluate the emission of PM	(17)
Emission efficiency NOx	Evaluating NOx emissions	(18)
Cost efficiency	Evaluate the cost of using different fuels	(19)

Source: Author (2016).

3.1.3 Eco-efficiency Assessment Procedure

In order to facilitate the adoption of the proposed method by companies, railways or regulatory agencies, a procedure for evaluating eco-efficiency was drawn up including a flowchart divided into five phases, with a total of 14 steps as shown in Figure 1:

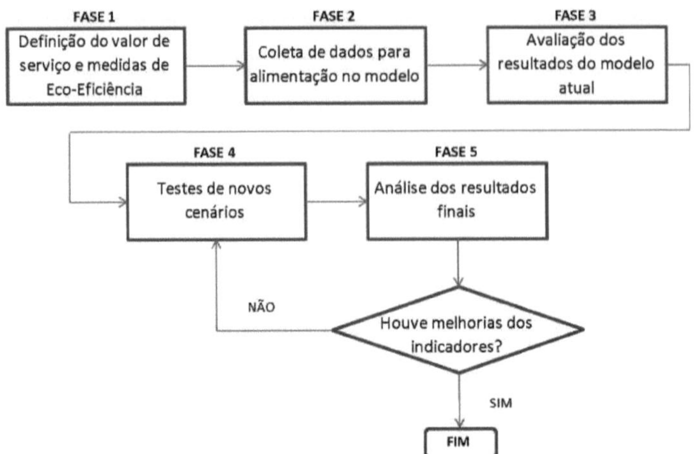

Figure 1 - Stages of the Eco-efficiency assessment procedure.
Source: Author (2016).

PHASE 1: Definition of service value and eco-efficiency measures

1) Definition of service value and environmental influence indicators;
2) Obtaining eco-efficiency measures;

PHASE 2: Collecting data to feed into the model

3) A survey of the fuels and technologies to be compared;
4) Survey of the fleet of locomotives used, their consumption and emission factors;
5) Literature review;
6) Data processing for insertion into spreadsheets;

PHASE 3: Evaluation of the results of the current model

7) Development of spreadsheets responsible for calculating eco-efficiency measures;
8) Inserting the processed data into the spreadsheets;
9) Obtaining and analysing eco-efficiency measures for the railway's current operating mode;

PHASE 4: Testing new scenarios

10) Creation of different operating scenarios with the use of alternative fuels;

11) Use the formulae set out in this chapter to compare the new scenarios with the current one;

12) Processing the data and obtaining the results for each additional scenario created;

PHASE 5: Analysing the final results

13) Evaluation and comparison of the results of the scenarios created in view of the current operating scenario.

14) Evaluate whether there have been any improvements compared to the current scenario. If so, propose implementing the scenario with the best results. If there is no scenario that is substantially better than the current operating mode, review the proposed scenarios and reprocess the data until scenarios with better results are obtained.

In the first stage of the first phase, the railway or agency must define which indicators are used for Service Value and Environmental Influence. For Service Value, a widely used indicator such as TKB is recommended, so that the results obtained can be compared with other railways. For Environmental Influence indicators, the railway or agency should assess which polluting gases are most important to analyse, taking into account local legislation: if there are emission regulations for specific gases, these should be included as indicators. Other factors should also be evaluated for the choice, such as the particular interests of the railway, local communities, or even national and international references.

During the third stage of the second phase, the company must select which alternative fuels and technologies will be considered in the study, in addition to collecting fuel data. Data such as calorific value, density, ignition point, purchase cost, among others, should be considered. The ease of access to new fuels should also be assessed, as well as whether their cost is higher or lower than that of the standard fuel, the need to adapt

the engines or other mechanical and electrical parts of the locomotives, the need for additional facilities to store the new fuels, among others. In the fourth stage, a detailed survey is carried out of the locomotive fleet, such as the number and model, whether there have been any modifications to the original design, the actual fuel consumption of each locomotive model, the duty cycle of each train formation and the pollutant emission factors for each locomotive model.
To obtain fuel consumption and emissions, a test bench should be used with locomotives from the existing fleet, where real data is collected using spreadsheets. At the same time, it is recommended that tests with the alternative fuels be carried out on part of the fleet, in an assisted manner, driving regular trains along the same lines as the standard trains that run on the lines. This aims to check the functionality of the new fuels, the assimilation of the particularities of operation, whether there has been a change in the way the trains are driven as a result of the change - such as loss of power, change in time to complete the same journey when compared to traditional fuel, among other factors that may not have been taken into account previously. The fifth stage includes research into the emissions of similar national and international railways. Railways in countries such as the USA publish their emissions, making it easier to access data for comparison purposes, as do several European railways. The sixth stage involves processing the data for insertion into the spreadsheets.

The seventh stage involves developing spreadsheets to automate the calculations based on the input data, using programmes such as Microsoft Excel. In the eighth stage, the data is entered into the previously prepared spreadsheets. The data is then processed in order to obtain the results for each of the Eco-efficiency Measures for the current operating model. In the ninth stage, the data obtained for the current model is analysed and can be compared with results obtained by other railways from the same or other companies, or even from other countries. This stage provides a "snapshot" of the current operating model, used as a reference by the company.

In the tenth stage, new scenarios are proposed that include the use of alternative fuels and technologies. The mix of fuels considered must take into account factors such as: what is the greatest possible blend of an alternative fuel that at the same time requires the least modifications to the assets, what combination results in the lowest emissions without the need for large investments, what associated fuels can reduce operating costs, i.e. have a lower acquisition cost than the standard fuel. A mixture of two fuels can also be taken into account, one of which is more expensive and the other cheaper than the standard fuel, as long as it results in lower emissions or greater energy efficiency. In the eleventh stage, the data collected for the new fuels is entered into the formulas proposed in this chapter, making it possible to generate values on the same basis as the railway's standard scenario. In the twelfth stage, the calculations are carried out using the formulae in the spreadsheets and an assessment is made as to whether there is any improvement. If so, the indicators of the proposed scenarios are identified as having improved compared to the railway's standard operating mode.

In the thirteenth and fourteenth stages, the results obtained for the proposed scenarios are analysed in comparison to the railway's current operating scenario, i.e. whether or not there is a consistent improvement in various Eco-Efficiency Measures. If not, new scenarios should be proposed so that the results of at least some of them are superior to the standard operating model.

If consistently better results than the current one have been obtained, the results should be normalised. The normalisation scoring method should take into account results from proposed scenarios that are better than the current scenario, as well as emission limits adopted as a reference based on local or international legislation, or even emissions from other railways. 1 point should be awarded for each measure in each of the scenarios that fulfils the established requirements, and 0 points otherwise. Then all the points obtained by each of the scenarios must be added together. The scenario with the highest score is considered the best, which would indicate a possible scenario to be adopted by the railway.

It is strongly recommended that a sensitivity analysis be carried out for the Cost Efficiency measure, since variations in the purchase price of alternative fuels can generate significant changes in the results, making the switch to a cleaner fuel financially unsustainable. For this analysis, the price of all fuels should be fluctuated by a percentage above and below the current market value. The percentage for variation should be defined by the company with a view to credible levels for price fluctuation over a horizon that covers the estimated period for financial return. If, after the sensitivity analysis, the results indicate that there are proposed scenarios with gains compared to the current scenario, the scenario with the greatest gain should be chosen, or even the scenario with gains in strategic measures for the company, such as a reduction in the emission of a certain pollutant, or a reduction in the total cost of operation.

Once the scenario to be adopted has been chosen, the measures needed to implement it must be assessed, such as the need to adapt the company's assets and industrial park, hiring labour, new maintenance and operating procedures, carrying out works and their impact on local communities, the space available in the event of the need to expand infrastructure, the logistics needed to transport, receive and supply the new fuels, the impact on suppliers or local communities caused by a reduction in the purchase of the railway's standard fuel once the new fuels come into play, among others.

CHAPTER 4

FUELS APPLICABLE TO LOCOMOTIVES

In addition to traditional diesel oil, there are several alternative fuel options for the locomotives on the railway studied, and only those that were less aggressive to the environment were selected. The choice should be prioritised by fuels that require less intervention in the locomotive engines, easier refuelling, greater emission reductions, lower purchase costs and technology that has been tried and tested around the world. The following is a brief description of the fuels: soya biodiesel, liquefied natural gas and petroleum diesel.

4.1 BIODIESEL

Biodiesel is a clean-burning fuel derived from natural and renewable sources such as plants. Like petroleum-based diesel, biodiesel operates in ignition-combustion engines. The use of biodiesel in conventional diesel engines results in a substantial reduction in hydrocarbons, carbon monoxide and particulate matter.

Among its advantages, it is a fuel that is simple to use, biodegradable, non-toxic and essentially free of sulphur and aromatic compounds. Table 3 shows the percentage of biodiesel used in Brazil over the years. The nomenclature adopted is as follows: the letter B indicates the presence of biodiesel and the preceding number the percentage of this fuel in the mixture, i.e. B6 indicates 6% biodiesel and 94% diesel.

Table 3 - Evolution of the percentage of biodiesel added to diesel in Brazil

Year	Percentage of Biodiesel blended with Diesel
2005 to June 2008	2%
July/2008 to June/2009	3%
July/2009 to December 2009	4%
January/2010 to June/2014	5%
July/2014 to October 2014	6%
November 2014 onwards	7%

Source: Petrobras (2015).

The following advantages of using biodiesel can be mentioned:

-It is a renewable energy source;

-It is carbon neutral and is neutralised by plants;

-Contributes to job creation in the primary sector of the economy, in order to foster the development of the Brazilian government;

-Environmental gain for the whole planet with the reduction of polluting gases;

-The viability of direct use was proven by the evaluation of the engine components, which showed no residues that compromised performance;

-It is a clean and renewable source of energy.

4. 2LIQUEFIED NATURAL GAS

Liquefied natural gas (LNG) consists mainly of methane (CH_4), with a minimum concentration of 85 per cent, as required by the ANP for the south-east of the country. Natural gas originates from the remains of plants and animals that have been subjected to great pressures imposed by an extensive layer of soil and rock over them for centuries. Its advantage is that it has lower atmospheric emissions than diesel oil,

mainly nitrogen oxides (NO_X), and it is currently marketed at a lower cost.

4.3 DIESEL OIL

The alternative fuels tested were compared with diesel oil, the latter being derived from petroleum and made up mainly of hydrocarbons with low concentrations of sulphur, oxygen and nitrogen. It is produced by the process of fractional distillation of petroleum, and the National Petroleum, Natural Gas and Biofuels Agency (ANP) is responsible for specifying its characteristics in Brazil. In Brazil, it currently requires that 7% biodiesel be added to ordinary diesel oil, which is the standard fuel used by locomotives.

4.4 CHARACTERISTICS OF THE FUELS TESTED

Below are the calorific values, emission factors and costs of each fuel tested on the railway.

Table 4 - Calorific value, emission factors and fuel costs

	B5	LNG*	B25
Calorific value	37,711 kJ/l	40,039 kJ/m^3	37,065 kJ/l
CO_2 emission factor	2.70 kg/l	2.03 kg/m^3	2.65 kg/l
NOx emission factor	44.3 g/l	10.09 g/m^3	45.23 g/l
PM emission factor	1.62 g/l	0.14 g/m^3	1.46 g/l
CO emission factor	4.5 g/l	5.9 g/m^3	4.05 g/l
Fuel costs	1.09 US$/l	0.69 US$/m^3	1.15 US$/l

* vaporised

Source: EFVM (2013)

It can be seen that B25 has a higher NOx emission factor than LNG or B5. On the other hand, it generates less CO_2, CO and PM. LNG emits less CO_2, less than 25 per cent of NOx and virtually zero PM, but it emits more CO than B5 or B25. Each cubic metre of LNG has more calorific value than a litre of B5 or B25, which means it generates more energy during combustion. The emission figures in Table 4 were obtained from tests carried out on locomotives similar to those in the railway's fleet.

CHAPTER 5

ATMOSPHERIC POLLUTANTS

The pollution produced by any motor vehicle comes mainly from the combustion process of burning fuels, with a smaller contribution from tyre wear, burning and disposal of lubricating oil, the vehicle manufacturing process, etc. In a perfect engine, the oxygen in the air would convert the hydrogen in the fuel into water and the carbon into carbon dioxide, as shown in Figure 2. However, the process is not perfect and ends up emitting various pollutants (MARTINS, 2006).

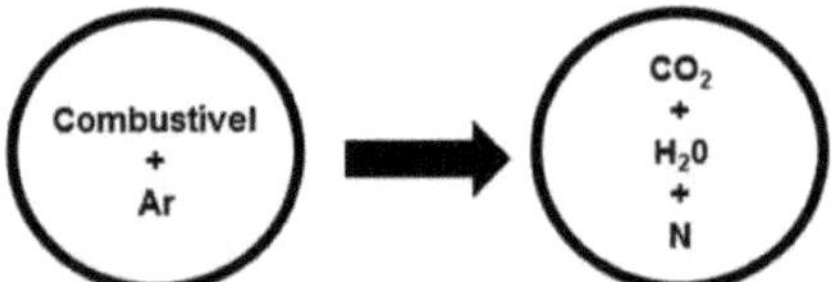

Figure 2 - Perfect combustion process in an internal combustion engine
Source: Martins (2006, p. 201).

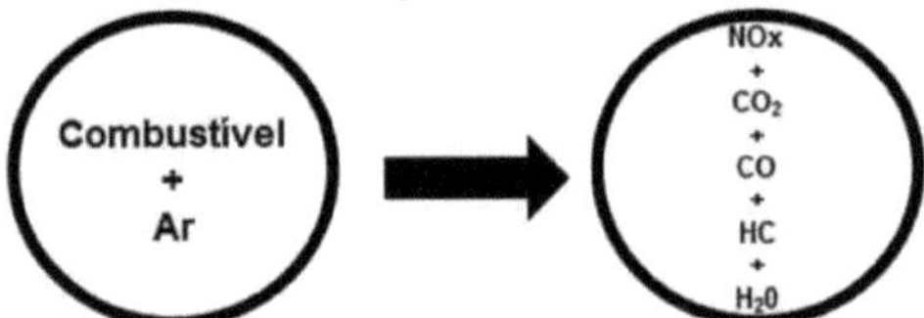

Figure 3- Real combustion process in an internal combustion engine
Source: Martins (2006, p. 201).

Despite all the technological development aimed at minimising emissions, the number of vehicles and the distance travelled have grown dramatically over the years and, even with the changes in the sector, the process of reducing emissions has not been effectively controlled. As a result, the level of reduction

of various pollutants has been modest, with the exception of lead (around 95 per cent), due to the fact that it was banned from fuels at the end of the last century (MOREIRA, 2012). In order to better understand how each of these pollutants harms health and the environment, we will briefly explain their characteristics.

5.1CARBON MONOXIDE (CO)

CO is an odourless and colourless gas, formed both in the process of burning fuels, heating systems, coal-fired power stations, burning biomass and tobacco, and in natural sources such as volcanic activity, electrical discharges and natural gas emissions (Ministry of the Environment, 2014).

The main route of exposure to carbon monoxide is respiratory, so acute poisoning can be fatal to humans. When inhaled, the gas is rapidly absorbed by the lungs and already in the circulatory system it binds with haemoglobin in a stable way. It thus prevents oxygen transport, causing tissue hypoxia, i.e. a low concentration of oxygen in organic tissues. At low concentrations, it can cause fatigue and chest pains (CETESB, 2014).

5.2NITROGEN OXIDE (NOX)

NOx is a highly oxidising pollutant gas that is also extremely important in the formation of ozone in the troposphere. In addition to its effects on human health, this type of pollutant also influences global climate change. Sources of NOx include locomotives, passenger cars, internal combustion engines, thermoelectric power stations and steel mills. Natural sources include forest fires, heat generated by lightning and microbial activity in soils (Ministry of the Environment, 2014).

With regard to human health, the adverse effects of exposure at high levels are difficulty breathing, changes in lung defence and aggravation of respiratory and cardiovascular diseases. It also increases susceptibility to asthma and bronchitis, especially in children, the elderly and at-risk groups (CETESB, 2014).

In the environment, when nitrogen oxide is oxidised in the atmosphere, it produces nitric acid (HNO_3), which is responsible for increasing the acidity of rain. When acid rain hits the surface, it alters the chemical composition of the soil and water, affects food chains, destroys forests and crops, corrodes metal structures, among other things (ECYCLE, 2014).

5.3PARTICULATE MATTER (MP)

PM can be defined as a complex set of small diameter pollutants made up of smoke, dust and any type of solid and liquid material that is suspended in the atmosphere due to its small size. The main sources are the burning of fossil fuels, the burning of plant biomass, ammonia emissions in agriculture and emissions from roadworks. PM can be subdivided into (ECYCLE, 2014):

- Total Suspended Particles (TSP): these are particles whose aerodynamic diameters are smaller than 50 iim. so that part of this material can be inhaled and cause health problems;

- Final Inhalable Particles ($PM_{2.5}$): are those with a diameter smaller than 2.5 iim. Due to their small size, they can penetrate deep into the respiratory system, reaching the pulmonary alveoli;
- Smoke (FMC): directly related to the combustion process of fossil fuels.

5.4GREENHOUSE GASES (GEE)

The two main sources of GHG emissions are the burning of fossil fuels and the deforestation of tropical forest regions. The main effects of GHGs are an increase in global temperature, which can cause the polar ice caps to melt, sea levels to rise, acceleration of the desertification process, imbalance in ecosystems and changes in the course of sea currents, among others. The main greenhouse gases are (IPAM, 2014):

- Carbon dioxide (CO_2): it remains in the atmosphere for at least a hundred years and its main sources of emissions are the burning of fossil fuels, forest fires and deforestation (responsible for the extinction of CO_2 absorption in the air by plants);
- Methane (CH_4): primary component of natural gas, also generated by the digestive

system of livestock, landfills, mining, among others;

- Nitrous oxide (N_2O): gas released by microorganisms in the soil (during the nitrification process) and increasingly released into the air due to the use of fertilisers;
- Chlorofluorocarbons (CFCs): widely used in air conditioners and refrigerators.

5.5POLLUTANT EMISSION LIMITS

As there are no pollutant emission standards for locomotives in Brazil, international references must be used, with US and European legislation being the most commonly referenced.

The United States Environmental Protection Agency (EPA) is a US government agency that sets emissions standards for on- and off-road vehicles. All American railways must meet the standards imposed by the EPA, and these have been published in the US Code of Federal Regulations (eCFR) - Chapter 40, Part 89 (DIESELNET, 2016; ECFR, 2016). The EPA's standards are to some degree harmonised with the European emissions standard for off-road vehicles, and the European model is also divided into four stages. Stage 4 is the most restrictive and has been in force since 2014. Stage 5, which is even more restrictive than Stage 4, was proposed in 2014 but has not yet come into force (EPA, 2012). The European standard is measured in grams per kilowatt hour, while the American standard is measured in grams per HP hour. When brought into the same unit, the American standard is more restrictive for all the pollutants referenced in this book (EPA, 2016; DIESELNET, 2016). Therefore, because they are more restrictive than the European standard and American railways focus on freight transport while European railways focus on passenger transport, the EPA standard was adopted.

The EPA defines five standards, called TIERs. TIER 0 is the standard that allows the highest emissions and TIER 4 is the most restrictive, allowing the least pollution (EPA, 2012). The US government defined that from January 2015 all locomotives should meet the requirements of the TIER 4 emissions standard. Table 5 shows the emission standards for each TIER for locomotives that run on the trunk line, i.e. not exclusively for shunting, and emissions are measured in grams per HP-hour. In order to compare the emissions standard defined by the EPA with the emissions generated by the railway, both are converted to the same unit of measurement.

Table 5 - Emission standard for locomotives defined by the EPA

Cycle	**TIER**	**NOx (g/HP-hr)**	**PM (g/HP-hr)**	**CO (g/HP-hr)**
	0	9,50	0,22	5,00
	1	7,40	0,22	2,20
o Q ti 2 H ti	2	5,50	0,10	1,50
	3	5,50	0,10	1,50
	4	1,30	0,03	1,50

Source: Adapted from EPA (2016).

CHAPTER 6

PRACTICAL APPLICATION OF ECO-EFFICIENCY CALCULATIONS ON RAILWAYS

In order to demonstrate how eco-efficiency evaluation is an important factor in decision-making, both from an environmental and financial point of view, this book presents a real-life application of the method. The method presented above was applied to the Vitória-Minas Railway (EFVM), which is one of the most important railways in the country. It transports more than 120 million tonnes of products a year along 905 kilometres of track. It has a fleet of 322 locomotives and approximately 14,000 wagons that daily form 30 high-load trains per axle, pulled by up to three locomotives and 252 ore wagons weighing 110 tonnes each. Its lines cross two states: Minas Gerais and Espírito Santo, and a complete journey cycle (port-to-port) takes around three days depending on the destination mine. The railway reaches 18 mines in the state of Minas Gerais, and its route can be seen in Figure 4.

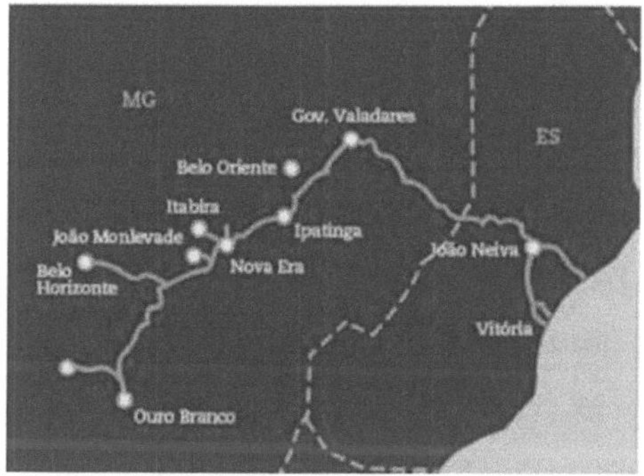

Figure 4 - Vitória-Minas railway.
Source: EFVM (2015).

Currently, the entire locomotive fleet uses B7 diesel oil as fuel, i.e. 7% biodiesel added to regular diesel oil, but as B5 was still in force in the country during the tests, the performance of the alternative fuels was compared to it.

The fuels tested were: B5, B25, natural gas combined with B5 and finally natural gas + B25. The emissions data was obtained through practical tests with railway locomotives, following the specifications of the United States Environmental Protection Agency (EPA), the US agency responsible for regulating emissions standards on railways in the United States.

A typical EFVM locomotive - model BB40 - used in the tests can be seen in Figure 5. This model has a power of 4,000 HP, generating energy of up to 3,100 kWh, and is capable of pulling 84 ore wagons weighing 110 tonnes each along the railway lines.

Figure 5 - BB40 locomotive used in the tests.
Source: Author (2015).

B25 was tested on part of EFVM's fleet between 2008 and 2014, being purchased in its pure form, 100% biodiesel (B100), and then mixed with B5 until a mixture of 25% biodiesel and 75% ordinary diesel was achieved. As a result of using B25, some maintenance plans had to be carried out more frequently on the locomotives. Seals, hoses and o-rings had to be replaced with new models to avoid the wear and tear caused by biodiesel. As the calorific value of biodiesel is lower than that of diesel, an increase in consumption of around 2.5 per cent was observed during operation, but in general, operation with B25 followed the same patterns as operation with diesel oil, and there were no differences in how the train was driven by the drivers. Blends with a higher biodiesel content were not tested because they could compromise the operation of the engines.

The LNG tests began in 2008 and were completed in 2015. They consisted of a technical and economic assessment of the use of natural gas mixed with diesel in locomotives, taking into account factors such as power, consumption and pollutant emissions, as well as analysing the useful life of components.

In order to use LNG as a fuel, locomotives need to have their engines converted using so-called dual-fuel kits. These kits allow the engine to run on a mixture of natural gas and diesel simultaneously, with no loss of power, unlike traditional passenger cars. This is due to the fact that during operation there is always a portion of diesel mixed with the gas, since the engine's compression ratio is not sufficient to cause the gas to combust. As there is no spark plug, when the diesel ignites it causes the gas to ignite, thus obtaining full engine power. The proportion of gas mixed with diesel is set by a computerised system and can vary from 30% to 80% depending on conditions such as the power required for traction, temperature and gas pressure.

A tank wagon also needs to be coupled to the locomotives to transport the natural gas. Inside the tank, the product is in the liquid phase and is vaporised by heat exchangers in the wagon before being transported to the locomotives. Once in the locomotive, the gas has its temperature and pressure regulated for admission to the engine, thus mixing with the B5. The railway has two wagons designed for this purpose called TGE (numbers 296.7014 and 296.700-6), Figure 6. The TGE wagons have a storage capacity of 38 m^3 of LNG, which is equivalent to 18,000 litres of diesel - once liquefied, the gas occupies a volume 580 times smaller than the gaseous state.

Figure 6: View of the LNG wagon (TGE)
Source: Author (2015).

TGE wagons are designed to power two locomotives simultaneously during journeys, operating either between or behind the locomotives at the head of the train, as shown in the figure:

Figure 7: Possible formations between locomotives and LNG wagon
Source: Author (2015).

More than 50 regular journeys with the modified locomotives hauling iron ore trains were carried out and monitored, totalling more than 100,000 km and 1,700 hours of operation, through which operating data was collected. Static tests were also carried out to collect data on consumption, power, current, voltage, ambient and coolant temperature, diesel oil pressure, among others. The power measurements showed that there were no significant changes between operation in diesel mode and dual-fuel mode.

6.1 PROPOSED SCENARIOS

Various scenarios were created to assess the impacts of rail transport on the EFVM.

Table 6 shows the five scenarios analysed.

Table 6 - Summary of scenarios

Scenario	Description
Scenario 1	All locomotives use B5 (current operation)
Scenario 2	All GE BB40 locomotives use LNG and B5, the others B5
Scenario 3	All GE BB40 locomotives use LNG and B5, the others B25
Scenario 4	All GE BB40 locomotives use LNG and B25, the others B25
Scenario 5	All locomotives use B25

Source: Author (2016).

The first scenario portrays the reality of current railway operations at EFVM. The other scenarios were evaluated based on the use of alternative fuels in various proportions.

Scenario 2 was proposed with the aim of studying the isolated effect of using LNG in BB40 locomotives, the model responsible for transporting iron ore on the railway, which has the most traffic. In this way, the investments for converting the engines would be concentrated only on the locomotives with the most traffic, thus avoiding capital expenditure on shunting locomotives.

Scenario 3 combines the effect of Scenario 2 with the use of B25 in the other locomotives, which have a lower running capacity. Scenario 4 uses B25 and associated LNG to maximise the use of alternative fuels. Scenario 5 considers using only B25 in all locomotive models, including those that run on trunk and shunting lines.

6.2 TOTAL ENERGY CONSUMPTION (E)

Figure 8 shows the result of total energy consumption (E) for each of the scenarios. Scenario 2 is the most efficient due to the use of LNG, and shows a small difference (0.06%) compared to Scenario 3. Scenario 5 is the least efficient because of the use of biodiesel, which increases fuel consumption due to its lower calorific value. Scenario 1 is better than scenario 5 (0.97%) since less biodiesel is used. The standard deviation of the results is 0.84 per cent, influenced by the consumption measurements taken for each type of fuel. This means that technically Scenarios 2, 3 and 4 are tied, with no statistically significant difference.

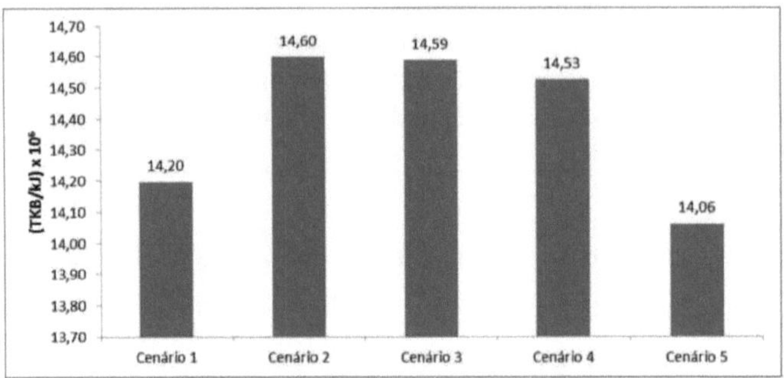

Figure 8 - Total Energy Consumption (E)

Source: Author (2016).

6.3 TOTAL RENEWABLE ENERGY CONSUMPTION (RE)

The use of renewable fuels - Figure 9 - leads to better environmental performance, with Scenario 5 being the most favourable because it uses the largest share of biodiesel. Scenario 4 comes in second, despite having a much lower result than Scenario 5. Scenario 2 has the lowest use of renewable energy due to the replacement of B5 with LNG, which is not considered a renewable fuel because it is derived from oil. Scenario 3 is better than Scenario 2 due to the use of B25 in the BB36 and DDM locomotives.

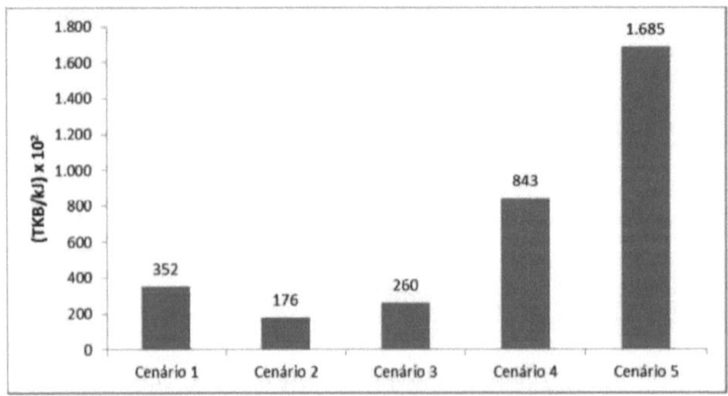

Figure 9 - Total Renewable Energy (RE) Consumption

Source: Author (2016).

6.4 CARBON DIOXIDE (CO_2) EMISSIONS

Scenario 4 shows the best result for CO_2 because LNG combined with biodiesel provides lower emissions. Scenarios 2 and 5 show fairly similar values and Scenario 1 shows the highest emissions. Since biodiesel is considered 100% renewable, CO_2 emissions can be reported separately from total emissions - Figure 10.

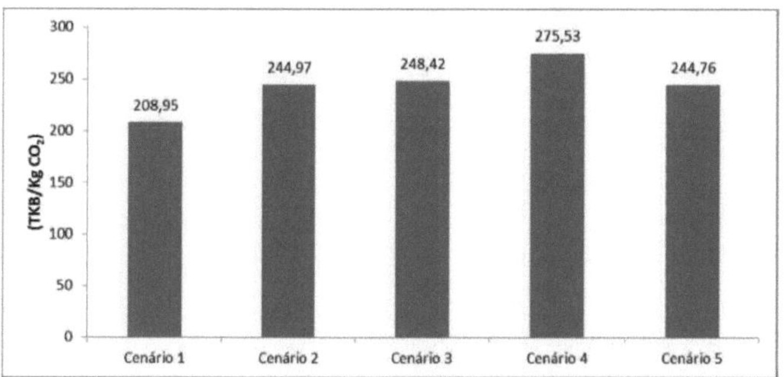

Figure 10 - Carbon dioxide emissions (CO)2
Source: Author (2016).

6.5 CARBON MONOXIDE (CO) EMISSIONS

Scenario 5 shows the best results, since CO emissions from biodiesel are lower than from B5 or LNG. Scenarios 2, 3 and 4 have higher CO emissions than Scenario 1, since LNG emits more carbon monoxide than B5 or B25. Scenarios 2 and 3 are tied, with a difference of 499 grams, i.e. within the margin of error of 1,871.14 grams. The differences in the other results were all significant. The lower horizontal line represents CO emissions for TIER 0 [16,436 TKB/CO (g)], the middle horizontal line for TIER 1 [37,356 TKB/CO (g)] and the upper horizontal line for TIERs 2, 3 and 4 [54,788 TKB/CO (g)]. As can be seen, all the scenarios meet the requirements for carbon monoxide emissions for TIER 4.

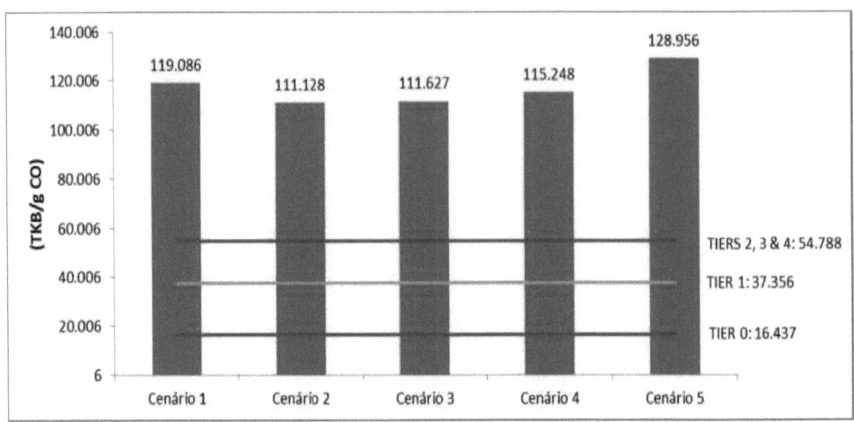

Figure 11 - Carbon Monoxide (CO) Emissions
Source: Author (2016).

6.6 PARTICULATE MATTER (MP) EMISSIONS

Scenario 4 shows the best results since the PM emissions from LNG and biodiesel are lower than those from petroleum diesel. Scenario 1 has the highest PM emissions, followed by Scenario 5. In order for the TIER 4 standard to be achieved, 2,739,421 TKB/MP (g) must be reached, so no scenario was able to achieve it. As can be seen, Scenarios 2, 3 and 4 meet the requirements of TIER 1, and Scenarios 1 and 5 are not even able to meet the minimum emission standards of TIER 0.

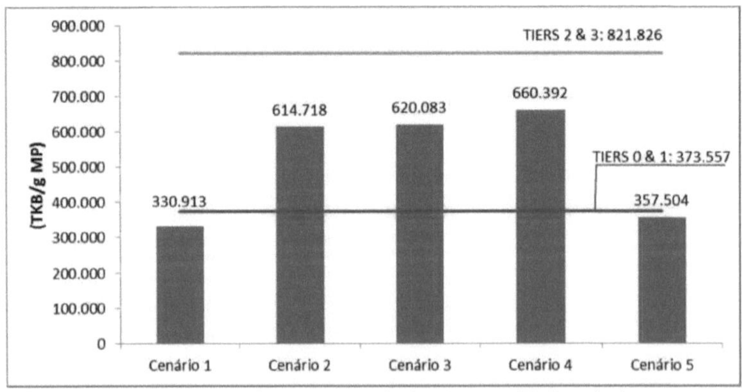

Figure 12 - Particulate Matter (PM) emissions
Source: Author (2016).

6.7emissions of nitrogen oxides (nox)

Scenario 2 has the lowest NOx emissions because LNG emissions are much lower than those of petroleum diesel or even biodiesel. Scenarios 3 and 4 show slightly lower emissions than Scenario 2 because biodiesel emits more NOx than petroleum diesel, but Scenario 3 shows a result within the margin of error of 10,456.33.

TKB/NOx (g) when compared to Scenario 2. Scenario 5 has the highest NOx emissions, even higher than Scenario 1, caused by the increase in fuel consumption caused by the use of biodiesel in locomotives, also associated with its higher emission factor. The TIER 4 standard could only be achieved with emissions of 63,217 TKB/NOx (g) and none of the scenarios is capable of achieving it. It can be seen that Scenarios 2, 3 and 4 are capable of achieving TIER 3 requirements and Scenarios 1 and 5 meet TIER 1 emissions standards.

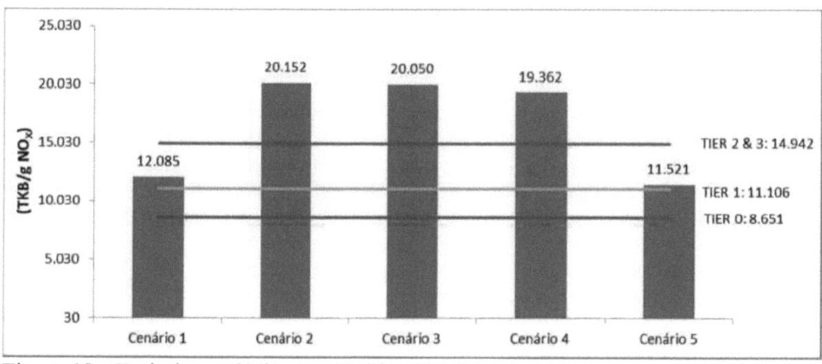

Figure 13 - Emissions of Nitrogen Oxides (NO)x
Source: Author (2016).

6.8 COST EFFICIENCY

As seen in Figure 14, Scenario 2 is the best in terms of costs, since LNG is cheaper than B25 or B5 in Brazil, but it is within the margin of error when compared to Scenario 3. Scenario 5 shows the worst result, as B25 is more expensive than B5 and also has a lower calorific value, increasing the locomotive's specific consumption. Scenario 1 is better than Scenario 5 because B5 costs less than B25.

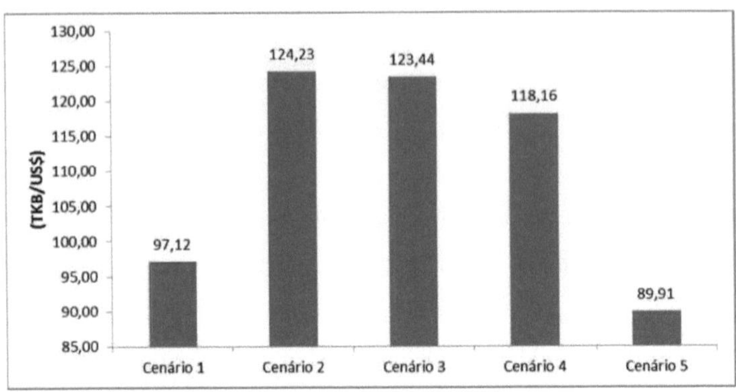

Figure 14 - Cost efficiency, considering only fuel costs.
Source: Author (2016).

Figure 15 takes into account, in addition to fuel costs, the installation and maintenance costs of dual-fuel kits for using LNG, as well as the extra maintenance required for locomotives using B25. The position of the scenarios remained the same as before, with Scenario 2 showing the best result, but with the lowest value. This means that even taking into account the maintenance and investment costs of replacing diesel with LNG and biodiesel, these alternatives are still better than the railway's current operating model.

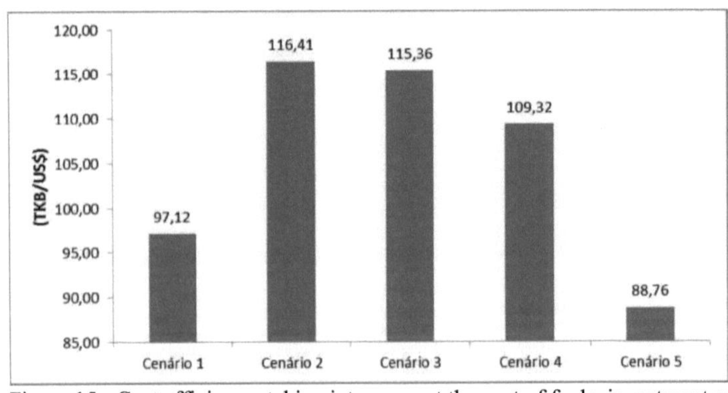

Figure 15 - Cost efficiency, taking into account the cost of fuels, investment and extra maintenance required for operation with LNG and B25.
Source: Author (2016).

6.9 GENERAL ANALYSIS

Scenario 1, the railway's current operating mode, does not show the best performance for any of the indicators, and the same is true for Scenario 3 - despite the latter showing 18.78 per cent lower costs. Scenario 2 shows the best results for EE, NOx and EC. Scenario 4 is best for CO_2 and PM emissions. Scenario 5 is the best for RE and CO. The company should therefore implement one of the three scenarios that best meets its needs, between Scenarios 2, 3 and 4, as they have lower costs than Scenario 1, mainly due to the large amount of LNG consumed. Even though biodiesel is more expensive than petroleum diesel, the large amount of LNG used in these scenarios makes up for it. Table 7 summarises the results, with the best highlighted in bold.

Table 7 - Summary of Results

Scenarios	E	RE	CO2	CO	PM	NOx	EC
Scenario 1	14,20	352	208,95	119.086	330.913	12.085	97,12
Scenario 2	**14,60**	176	244,97	111.128	614.718	**20.152**	**116,41**
Scenario 3	14,59	260	248,42	111.627	620.083	20.050	115,36
Scenario 4	14,53	843	**275,53**	115.248	**660.392**	19.362	109,32
Scenario 5	14,06	**1.685**	244,76	**128.956**	357.504	11.521	88,76

Source: Author (2016).

A sensitivity analysis of Cost Efficiency should also be carried out and considered fluctuations in fuel prices of -30 per cent and +30 per cent for B5, B25 and LNG.

Table 8 - Results of the sensitivity analysis (TKB/US$)

	Min.(-	**Curren t**	**Max (+30%)**
Scenario 1	74.71	97.12	138.75
Scenario 2	90.87	116.41	161.93
Scenario 3	90.52	115.36	158.98
Scenario 4	88.73	109.32	142.34
Scenario 5	68.48	88.76	126.11

Source: Author (2016).

In order to compare the results, all the values are placed on the same basis and the following criteria have been adopted:

-E , RE, CO_2, and CE: all scenarios that showed better results than Scenario 1 scored 1 point;

-CO : all the scenarios that reached the TIER 4 emission standard scored 1 point, since TIER 4 is the most restrictive emission standard in the United

States;

-PM and NOx: all scenarios that met the TIER 3 standard scored 1 point. Since the TIER 4 standard is very high in relation to all the scenarios analysed, the TIER 3 standard was adopted because it is the standard immediately below TIER 4.

Table 9 - Normalised results

Scenario	**E**	**RE**	CO_2	**CO**	**PM**	NOx	**EC**	**Total**
Scenario 1	0	0	0	1	0	0	0	**1**
Scenario 2	1	0	1	1	1	1	1	**6**
Scenario 3	1	0	1	1	1	1	1	**6**
Scenario 4	1	1	1	1	1	1	1	**7**
Scenario 5	0	1	1	1	0	0	0	**3**

Source: Author (2016).

Scenario 4 proved to be the best scenario for implementation on the railway. This scenario combines the use of natural gas and biodiesel, presenting lower operating costs than the current railway model and, at the same time, lower emissions, with the exception of CO emissions, but it still meets the requirements of the American TIER 4 standard, which is the most restrictive. If Scenario 4 is implemented, compared to Scenario 1 the railway could gain 2.3% in the energy efficiency indicator, 139% in the renewable energy indicator, 32% in CO_2 emissions, 100% in PM, 60.0% in NOx and 12.6% in Cost Efficiency. The only indicator that shows a worse result when comparing Scenario 4 to Scenario 1 is CO emissions, with a 3.2 per cent drop in the indicator, due to the use of LNG.

The method can be used as a support tool to help the railway decide on the best fuel mix and technology to use in order to minimise its environmental impact and operating costs.

CHAPTER 7

REFERENCES

ABDELAAL, M. M.; HEGAB, A. H. (2012). **Combustion and emission characteristics of a natural gas-fuelled diesel engine with EGR.** Energy conversion and management, 64, 301-312.

ALVES, F.; FIORIO, V. **How is ethanol produced?** Available at: <http://www.industriahoje.com.br/como-e-produzido-o-etanol>. Accessed on 15 December 2014.

ANP, National Agency for Petroleum, Natural Gas and Biofuels (2014). **Institutional folder**. Available at: http://www.anp.gov.br/?dw=2447. Accessed: 18/07/2014 at 16:52hs.

ANP, National Agency of Petroleum, Natural Gas and Biofuels (2014) **Monthly Biodiesel Bulletin**, April 2014. Available at: http://www.anp.gov.br/?dw=70554. Accessed: 01/08/2014 at 09:48hs.

TM-131 GAS ANALYSER. Available at: http://www.tecnomotor.com.br/novo/index.php/produtos/2015-04-02-20-05- 00/tm-131-gas-analyser. Accessed: 24/04/2014 at 18:52hs.

ANTF, National Association of Railway Transport, 2014, **Balance of the Rail transport 2013**. Available at: http://www.antf.org.br/index.php/noticias/5139-balanco-do-transporte- ferroviario-2013. Accessed: 16/07/2014 at 17:20.

ARB, California Environmental Protection Agency, Air Resources Board, (2014). Available at: http://www.arb.ca.gov/homepage.htm. Accessed: 03/09/2014 at 20:25hs.

BNDES (National Bank for Economic and Social Development). **Brochure on the Environment Line. National Bank for Economic and Social Development**. 2007.

CALLENBACH, E. **Ecological management eco management: the Elmwool Institute's guide to ecological auditing and sustainable business**. 1 ed. São Paulo: Cultrix, 1993.

CARRARETTO, C.; MACOR, A.; MIRANDOLA, A.; STOPPATO, A.; TONON, S. (2004). **Biodiesel as alternative fuel: experimental analysis and energetic evaluations.** Energy, 29(12), 2195-2211.

CEBDS (Brazilian Business Council for Sustainable Development). **Eco-efficiency**, Gazeta Mercantil, 10/08/2005, available at:<http://www.cebds.org.br/cebds/artigos.asp?area=>. Accessed on 10 December 2014.

CETESB (Environmental Company of the State of São Paulo). **Vehicle Emissions**

in the State of São Paulo - 2012. Available at: < http://www.cetesb.sp.gov.br/userfiles/file/ar/emissoes/relatorio-2012.pdf>. Accessed on 17 December 2014.

Available at :< https://www.dieselnet.com/standards/us/nonroad.php>. Accessed on 20 June 2016.

ELECTRONIC CODE OF FEDERAL REGULATIONS - ECFR. Available at: http://www.ecfr.gov/cgi-bin/text-idx?tpl=/ecfrbrowse/Title40/40cfr89_main_02.tpl>. Accessed on 20 June 2016.

Available at : <http://www.ecycle.com.br/component/content/article/35/1379-os-perigos-do-material-particulado.html>. Accessed on 08 January 2015.

Available at : <http://www.ecycle.com.br/component/content/article/63-meio-ambiente/2393-nitrogen dioxide-oxide-oxides-what-and-where-is-it-present-sources-cars-engines-burning-microbial-effects-health-bronchitis-respiratory-problems-autism-effects-on-the-environment-acid-smoke-alternatives-kitchen.html>. Accessed on 05 January 2015.

ENVIRONMENTAL PROTECTION AGENCY - EPA. Available at: http://www.epa.gov/otaq/standards/nonroad/locomotives.htm. Accessed June 2016.

HARRINGTON, H. J.; KNIGHT, A. **Implementing ISO 14000: how to update your EMS effectively.** 1 ed. São Paulo: Atlas, 2001.

HAY, W. W. (1982). **Railroad engineering** (Vol. 1). John Wiley & Sons.

HOLLIDAY, C.; SCHMIDHEINY, S.; WATTS, P. **Walking the Talk - The Business Case for Sustainable Development.** United Kingdom: Green leaf Publishing, 2002.

HOLT, D. J. (2004). **Alternative diesel fuels**. Fuel, 2014, 02-21.

IPAM (Amazon Environmental Research Institute). **What are the main sources of greenhouse gases from human activities**? Available at:<http://www.ipam.org.br/saiba-more/abc/mudancaspergunta/What-are-the-main-sources-of-greenhouse-gases-from-human-activities-11/3>. Accessed on 10 January 2015.

LEAL JUNIOR, I. C.; GARCIA, P. A. A.; D.'AGOSTO, M. A. **A data envelopment analysis approach to choose transport modes based on eco- efficiency**. Environment, Development and Sustainability, v. 14, n. 5, p. 767781, 2012.

LOBO, I. P.; FERREIRA, S. L. C.; CRUZ, R. S. **Biodiesel: quality parameters and analytical methods.** Química Nova, v. 32, n. 6, p. 1596-1608, 2009.

MARTINS, J. **Motores de combustão interna**. 2 ed. Porto: Editora Publindústria 2006.

MINISTRY OF THE ENVIRONMENT. **Atmospheric pollutants**. Available at: <http://www.mma.gov.br/cidades-sustentaveis/qualidade-do-ar/poluentes-

atmosf%C3%A9ricos#>. Accessed on 08 January 2015.

MOREIRA, S. M. S. R. **Influence of biodiesel on pollutant emissions from a turbo diesel engine.** 2012. 93f. Dissertation (Master's in Mechanical Engineering) - Faculty of Engineering, University of Porto, 2012.

MOTTA, R. A.; SILVA, P. C. M.; BRASIL, A. C. M. **Challenges of sustainable mobility in Brazil**. Revista dos Transportes Públicos, n. 34, p. 25-48, 2012.

OLIVEIRA, F. C. C.; SUAREZ, P. A. Z.; DOS SANTOS, W. L. P.. **Biodiesel: possibilities and challenges.** Redes, v. 1, n. 497, 2013.

PEREIRA, C. A. S. **Development of eco-efficiency indicators: application to the metalworking industry.** 2001. 121f. Dissertation (Master's in Environmental Engineering) - Faculty of Engineering, University of Porto, 2001.

PETROBRAS. Technical manual: Diesel S10. Available at: <http://www.sindipetro.com.br/arquivos/cartilhas/c0c98afc10ef12d086fedc8e07400707.pdf>. Accessed on 07 January 2015.

PETROBRAS. ÓleoDiesel . Disponívelem : <http://www.br.com.br/wps/portal/portalconteudo/produtos/automotivos/oleodiesel/>. Accessed on 20 December 2015.

PORTALBIODIESEL. Available at : <http://www.biodieselbr.com/biodiesel/biodiesel-brasil.htm>. Accessed on 10 December 2014.

SALGADO, V. G. **Eco-efficiency indicators and the transport of natural gas**. 1 ed. Rio de Janeiro: Interciência, 2007.

SANTANNA, I. **Emissions of polluting gases from diesel vehicles that cause environmental deterioration: a search for solutions to alleviate or control the problem**. 76f. 2012. Monograph (Degree in Environmental Engineering) - Minas Gerais School of Engineering. Minas Gerais, 2012.

SEIFFERT, M. E. B. **ISO 14001 environmental management systems: objective and economical implementation.** 1 ed. São Paulo: Atlas, 2005.

THE GREENHOUSE GAS PROTOCOL, Revised Edition. Chapter 9, **Reporting GHG Emissions**. World Resources Institute and World Business Council for Sustainable Development, March 2004, Title 40 of the U.S. Code of Federal Regulations (CFR), Part 92, Subpart B.

TIBOR, T.; FELDMAN, I. **ISO 14000: a guide to the new environmental management standards**. 2 ed.São Paulo: Futura, 1996.

UNCTAD, 2004, **A Manual for The Preparers and Users of Eco-Efficiency Indicators.** Available at:<http://www.unctad.org/en/docs/iteipc20037_en.pdf>. Accessed on 30 October 2014.

WBCSD; VERFAILLIE, H. A.; BIDWELL, R. **Measuring Eco-Efficiency. A Guide to Reporting Company Performance.** World Business Council for

Sustainable Development, 2000.

WORLD BUSINESS COUNCIL FOR SUSTAINABLE DEVELOPMENT, 2000. **Measuring Eco-Efficiency. A Guide to Reporting Company Performance**. World Business Council for Sustainable Development, Geneva.

Printed by Books on Demand GmbH, Norderstedt / Germany